Our solar system

	Diameter (in km)	Average distance from the Sun (in km)	Length of 'year' (in Earth-years)	Length of 'day' (in Earth-time)	Number of moons
Sun	1,392,000			27.5 days	0
Mercury	4,880	58,000,000	0.24	58.65 days	0
Venus	12,100	108,000,000	0.62	243 days	0
Earth	12,260	149,000,000	1	23hrs 56mins	1
Mars	6,790	228,000,000	1.88	24hrs 37mins	2
Jupiter	142,800	778,000,000	11.86	9hrs 50mins	16
Saturn	121,000	1,426,000,000	29.46	10hrs 14mins	23
Uranus	51,800	2,868,000,000	84.01	10hrs 49mins	5
Neptune	49,500	4,494,000,000	164.79	16hrs	2
Pluto	6,000	5,896,000,000	247.69	6.39 days	1

Illustrators
Colin Hadley title page, pages 8-9,
10-11, 14-15, 16-17, 22-23
Stephen Kyte pages 12-13, 20-21
Mark Longworth end sheets,
pages 24-25
Tom Stimpson cover,
pages 6-7, 18-19, 26-27, 28-29

Designer Pat Butterworth

First published 1984 by Walker Books Ltd,
17-19 Hanway House, Hanway Place,
London W1P 9DL

© 1984 Walker Books Ltd

First printed 1984
Printed and bound in Spain
by Artes Graficas Toledo, S.A. DL-TO-134-84

British Library Cataloguing in Publication Data
Boase, Wendy
Space traveller. – (Young explorers; 1)
1. Space flight–Juvenile literature
I. Title II. Series
629.4'1   TL 793

ISBN 0-7445-0110-5

Contents

In this book:
km = kilometres
kg = kilograms
°C = degrees Centigrade
hrs = hours
mins = minutes

To read the answer to a quiz, hold a mirror at the right-hand side of the words.

SPACE TRAVELLER

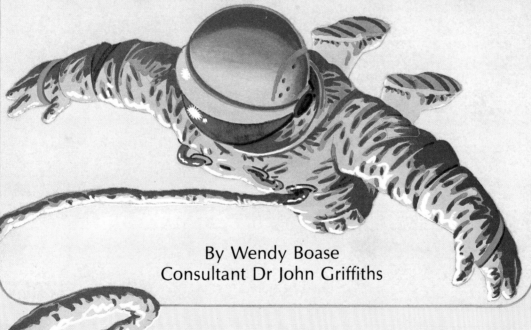

By Wendy Boase
Consultant Dr John Griffiths

WALKER BOOKS
LONDON

Sticking to Earth

Jump high! You'll always come down again. That's because an invisible force called gravity pulls you back to Earth. Gravity is like a very strong hand. If gravity didn't exist, you would float up into the sky. So would trees, houses, seas, bicycles and everything else. Suddenly cutting off gravity would cause our Sun and the nine planets in our solar system to fly off into space in all directions.

Earth is a special planet in our solar system because it has lots of water, a temperature that is ideal for life, and an atmosphere that we can breathe. Earth's atmosphere also stops the deadliest of the Sun's rays from getting through.

. .Earth's average distance from Sun 149 million km . . travels round Sun at 30km per secon

The great escape
To overcome gravity, a rocket must travel at 40,000 km per hour – about 20 times faster than Concorde. Looking down from high above Earth, you can see swirling clouds, orange desert, and vast areas of blue water. Water covers nearly three-quarters of our planet.

. . one moon . . . temperature extremes −88°C to 58°C . . . 24 hour day . . . .

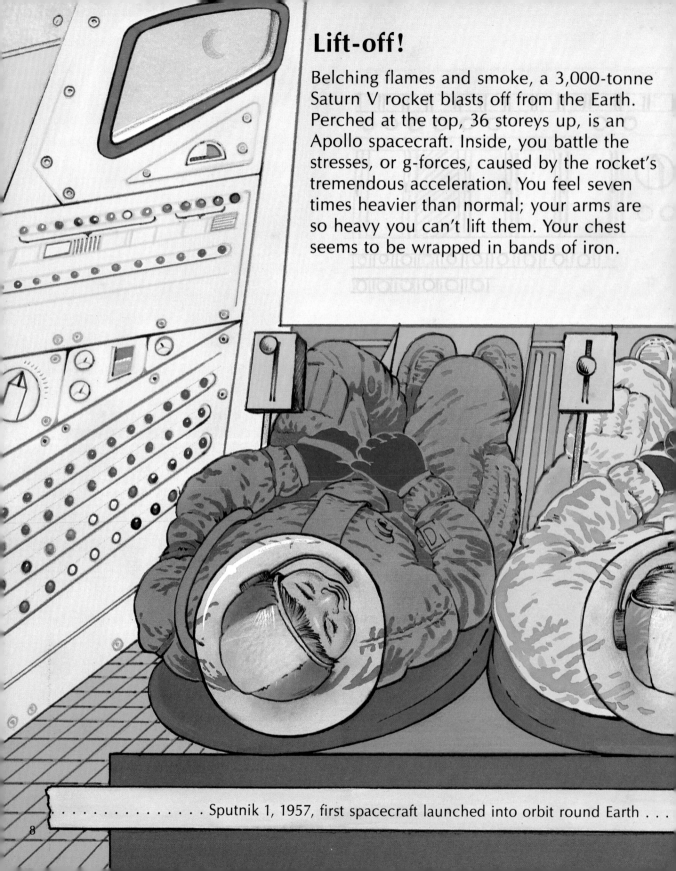

Lift-off!

Belching flames and smoke, a 3,000-tonne Saturn V rocket blasts off from the Earth. Perched at the top, 36 storeys up, is an Apollo spacecraft. Inside, you battle the stresses, or g-forces, caused by the rocket's tremendous acceleration. You feel seven times heavier than normal; your arms are so heavy you can't lift them. Your chest seems to be wrapped in bands of iron.

. . . . Sputnik 1, 1957, first spacecraft launched into orbit round Earth . . .

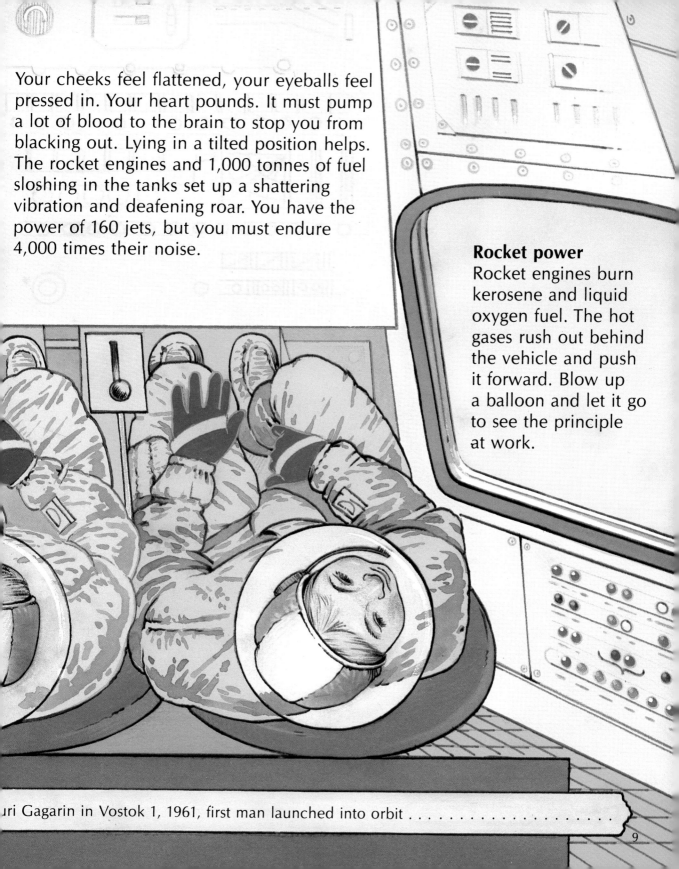

Your cheeks feel flattened, your eyeballs feel pressed in. Your heart pounds. It must pump a lot of blood to the brain to stop you from blacking out. Lying in a tilted position helps. The rocket engines and 1,000 tonnes of fuel sloshing in the tanks set up a shattering vibration and deafening roar. You have the power of 160 jets, but you must endure 4,000 times their noise.

Rocket power

Rocket engines burn kerosene and liquid oxygen fuel. The hot gases rush out behind the vehicle and push it forward. Blow up a balloon and let it go to see the principle at work.

...ri Gagarin in Vostok 1, 1961, first man launched into orbit . . . .

Topsy-turvy world

Empty space begins about 100 km above the Earth. You float because you are weightless, and there is no 'up' or 'down'. In an orbiting space station you have to adjust to a very strange environment. Doing even the simplest things can be difficult.

Walking normally is difficult, although you could easily turn a cartwheel! A gentle, swooping dive will carry you round the space station.

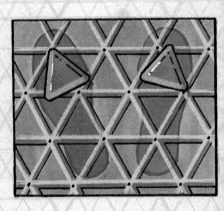

You need to exercise to keep fit in space. You can run round the walls like a fly, but you must move quite fast or you will float off them.

You have to anchor yourself to work or eat. Triangles on the soles of your shoes slot into the grids that make up floors and ceilings.

You shower in a bag, using a hand-spray and liquid soap. A vacuum hose sucks up floating water. The whole process takes about 45 minutes!

To sit on the toilet, you have your feet in straps and a belt across your lap. Air sucks waste from your body into bags which are dumped into space.

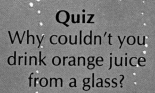

Quiz
Why couldn't you drink orange juice from a glass?

Answer
The orange juice would float away in little droplets.

On Earth, food for astronauts is dried, then frozen and put into plastic bags. You inject water to make the food swell, then suck it out.

The shrivelled food in the bags equals this Earth-meal. In space you can eat ham and chocolate pudding, but only sticky food will stay on a plate.

Your bed is a bag fixed to the wall. You wrap yourself in blankets, fasten a strap across your body, and pull the zip right up to your neck.

Floating about

If you jumped into space without a suit your blood would boil. (The higher up you go, even on Earth, the lower a liquid's boiling point.) A spacesuit also protects you from extreme temperatures. In space it can be 100°C in sunlight and −150°C in shade. A gold-plated plastic visor reflects the Sun's rays. A radio in your helmet links you to the space station, but your real life-line is the air-supply hose. If it broke, you would die in minutes, and drift off into the void.

Imagine wearing 14 layers of nylon, cloth, plastic and rubber, a helmet, boots 25 layers thick and gloves of woven metal. You would look gigantic and feel very hot! Pipes sewn into your underwear carry water to keep you cool. The water, and oxygen for breathing, are supplied through tubes. Other tubes carry away the carbon dioxide you breathe out. Sensors on your belt record your heartbeat and breathing rate.

Skylab

Space stations, or 'satellites', orbit the Earth just as our Moon does. They enable scientists to study space and to look back at Earth. Skylab was as big as a house. It orbited 500 km above Earth.

st space station put into orbit . . . Skylab, 1973, biggest space station put into orbit . . . . .

13

Lunar records

Go about 380,000 km past Skylab (the same distance as nearly 10 times round the Earth!) and you reach the Moon. Here, if you didn't have to wear a spacesuit, you could set some astonishing sports records. But choose the events carefully. Gravity is so weak that you bounce, so sprinting, for instance, would be very difficult!

Weightlifting

If you can lift a 30kg rock on Earth, you could lift a lunar rock weighing 180 kg. On the Moon everything weighs six times less than on Earth.

High jump

If you can jump a bar one metre high on Earth, you could clear six metres on the Moon. Gravity on the Moon is one-sixth as strong as on Earth.

Gymnastics

You can leap higher, and so stay up longer. If you are a good gymnast, you could turn about 10 full flips on the Moon.

. . . . Moon travels round Earth at 1km per second . . . diameter 3,476 km .

Quiz
Why couldn't you fly
a flag on the Moon?

Answer
Because there is no wind.
The Apollo Mission put up
a flag stiffened with a pole.

Throwing event
There is no air on the
Moon to slow down
flying objects. You
could throw a discus
further.

Kite flying
Don't go in for this
event. As there is no
air on the Moon, there
is no wind. Your kite
would be grounded.

Water events
Don't enter these
events. There is no
water on the Moon,
and you would not
enjoy swimming in
powdery dust!

erage temperature 120°C day, −180°C night . . Neil Armstrong, 1969, first man on Moon

Ask your computer

No one has yet gone beyond the Moon. If *you* want to travel further, you must provide your computer with a lot of extra information.

Feed the computer some facts on Venus (the planet closest to Earth) and Mercury (closest to the Sun) to see how you might survive on them.

INPUT: **Venus** is covered by dense clouds 30km thick.

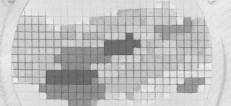

READOUT: Surface is rocky. Use radar to find landing site.

INPUT: Lightning bolts and sulphuric acid rain in clouds.

READOUT: Descend quickly, or acid will destroy spacecraft.

Quiz: Why would an ordinary thermometer be useless on Venus?
Answer: Because it measures only to 47°C.

INPUT: Atmosphere below cloud almost pure carbon dioxide. Light is lurid red.

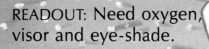

READOUT: Need oxygen, visor and eye-shade.

INPUT: Pressure same as 1km down in Earth sea. Surface 500°C

READOUT: Use a deep-sea type vehicle made of heat-resistant metal such as titanium.

INPUT: **Mercury,** smallest planet and fastest in orbit, races round Sun at 48km per second.

READOUT: Your rendezvous speed should be 28,000 km per hour.

INPUT: Closest to Sun; 350°C by day, −180°C at night.

READOUT: Wear a spacesuit coated in reflective material.

INPUT: Low gravity. Craters vast, mountains up to 4km high.

READOUT: Use the reversing retro-rockets to slow you down, and land on floor of crater.

INPUT: Which crater?

READOUT: Caloris Basin, 1,300 times bigger than the crater in Arizona. Ringed by mountains 2km high, so take care.

INPUT: Anything special?

READOUT: At night you see two brilliant lights: Venus (at 41 million km) and the blue Earth (at 91 million km).

Quiz
Why couldn't you have an ice-cream sundae on Mercury?

Answer
Not only would the ice-cream melt; so would the metal dish. (Tin melts at 230°C.)

Spot the difference

On this planet the Sun rises in the east and sets in the west, and the day is 24 hours long. Clouds float across the sky. The planet has two frozen poles, and regular seasons. In winter, frost covers the ground. There are deserts, winding valleys, extinct volcanoes and a huge canyon. Is it planet Earth? Could this scene be somewhere in Arizona?

Red planet

Mars is red, due to iron in the soil. Swirling red dust turns the sky pink. Martian winds can blow up to 500 km per hour. The worst hurricanes on Earth will gust up to only 350 km per hour.

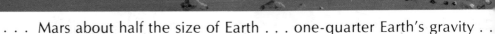
. . . . Mars about half the size of Earth . . . one-quarter Earth's gravity . .

Look at the pink sky, the small Sun and the two moons. This is Mars. The canyon, called the Valley of the Mariners, is 4,000 km long – the distance from London to Egypt! It wasn't carved by water, as Earth's Grand Canyon was, but by surface buckling. The volcano Olympus Mons, is three times as high as Mt Everest. Is Mars really like Earth?

Half-size Sun

The Sun is too far away to warm Mars. The temperature is about −30°C, falling lower at night. You would be alright in a well-insulated suit. It gets twice as cold as this in Siberia.

Thin atmosphere

You need to breathe oxygen on Mars. The atmosphere is mostly carbon dioxide, and is 100 times thinner than on Earth. That's like being about 32km up in Earth's atmosphere.

228 million km from Sun . . . two small moons . . . .

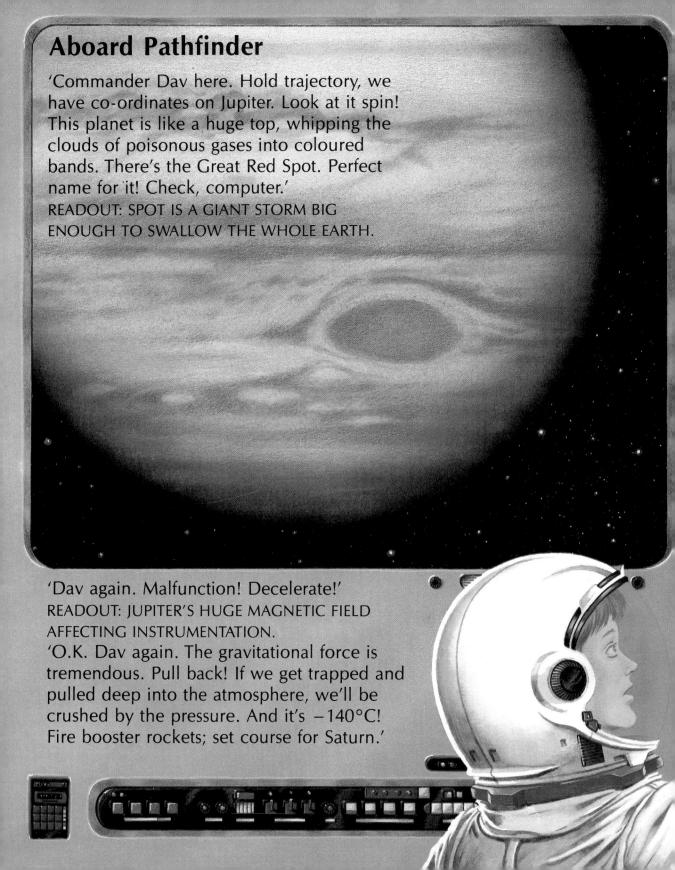

Aboard Pathfinder

'Commander Dav here. Hold trajectory, we have co-ordinates on Jupiter. Look at it spin! This planet is like a huge top, whipping the clouds of poisonous gases into coloured bands. There's the Great Red Spot. Perfect name for it! Check, computer.'
READOUT: SPOT IS A GIANT STORM BIG ENOUGH TO SWALLOW THE WHOLE EARTH.

'Dav again. Malfunction! Decelerate!'
READOUT: JUPITER'S HUGE MAGNETIC FIELD AFFECTING INSTRUMENTATION.
'O.K. Dav again. The gravitational force is tremendous. Pull back! If we get trapped and pulled deep into the atmosphere, we'll be crushed by the pressure. And it's −140°C! Fire booster rockets; set course for Saturn.'

'Commander Dav. I see the brilliant white arc of Saturn's rings. Decelerate! If we fly into that ring system at the wrong angle, Pathfinder will be riddled with holes. Edge-on. Go! Snowballs and icy hailstones are spinning by us. Look out! A big one. We're through! Check, computer!'
READOUT: THOUSANDS OF RINGS, ALL MADE OF BILLIONS OF LUMPS OF ICE AND ROCK.

'Dav. Viewing the surface of Saturn now. It's a ball of yellow cloud and gas at −170°C. No hope of landing. Check!'
READOUT: SATURN WOULD FLOAT ON WATER IF YOU COULD FIND A BIG ENOUGH OCEAN.
'What a sense of humour that computer has! Let's go! it's still 5,000 million km to the edge of our solar system.'

21

Empty highways

If you continued with Commander Dav, you would eventually reach the three outer planets in our solar system: Uranus, Neptune and Pluto. You might pass the odd asteroid, perhaps even a comet, but mostly you would see vast, empty black space. Here's the radio information relayed in words and pictures by a friend of yours at Mission Control.

Uranus
This planet is a giant green ball of hydrogen, helium and methane gases. Very poisonous, so don't stick your head out the window! Rings of gas, dust and rock orbit the planet, as well as five moons.

Neptune
A gas giant, very similar to Uranus. A hazy covering of yellow cloud hangs over it. Two moons.

Pluto
You would never have a birthday on Pluto. This frozen little planet orbits the Sun every 248 Earth-years! (Earth orbits every 365 days.) On Pluto it is −230°C. Pluto is the coldest planet in our entire solar system.

Uranus orbits Sun every 84 years . . . temperature −210°C . . . Neptune orbits Sun ever

Asteroids
Most asteroids orbit in a belt between Mars and Jupiter, but there are some of these tumbling chunks of rock out where you are. Their gravity is so weak that if you jumped, you would fly off into space.

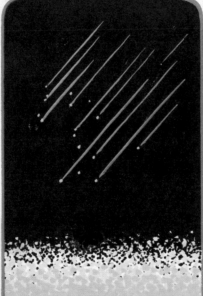

Meteoroids
These tiny grains of matter can travel faster than a rifle bullet. They could punch a hole in your craft.

Comets
Some of these giant dirty snowballs may sweep by. They get tails only if they go close to the Sun. The tail may be millions of kilometres long, but it could fit into a suitcase, because it's all dust and gas.

165 years . . . temperature −220°C . . . Pluto 5,896 million km from Sun . . . .

23

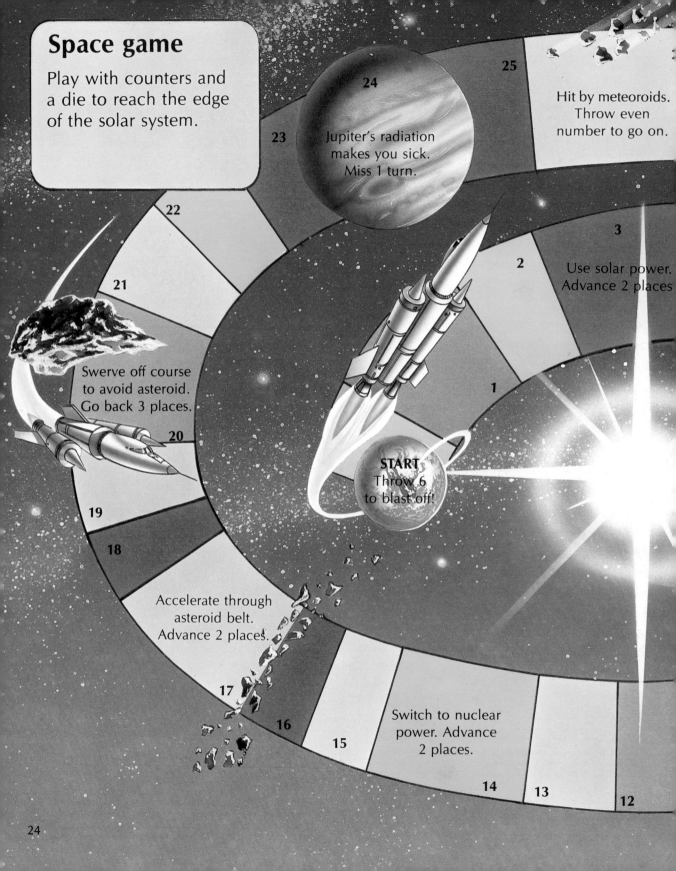

Space game

Play with counters and a die to reach the edge of the solar system.

24

23 Jupiter's radiation makes you sick. Miss 1 turn.

25 Hit by meteoroids. Throw even number to go on.

22

21

3

2 Use solar power. Advance 2 places

1

Swerve off course to avoid asteroid. Go back 3 places.

20

START Throw 6 to blast off!

19

18

Accelerate through asteroid belt. Advance 2 places.

17

16

15 Switch to nuclear power. Advance 2 places.

14 **13** **12**

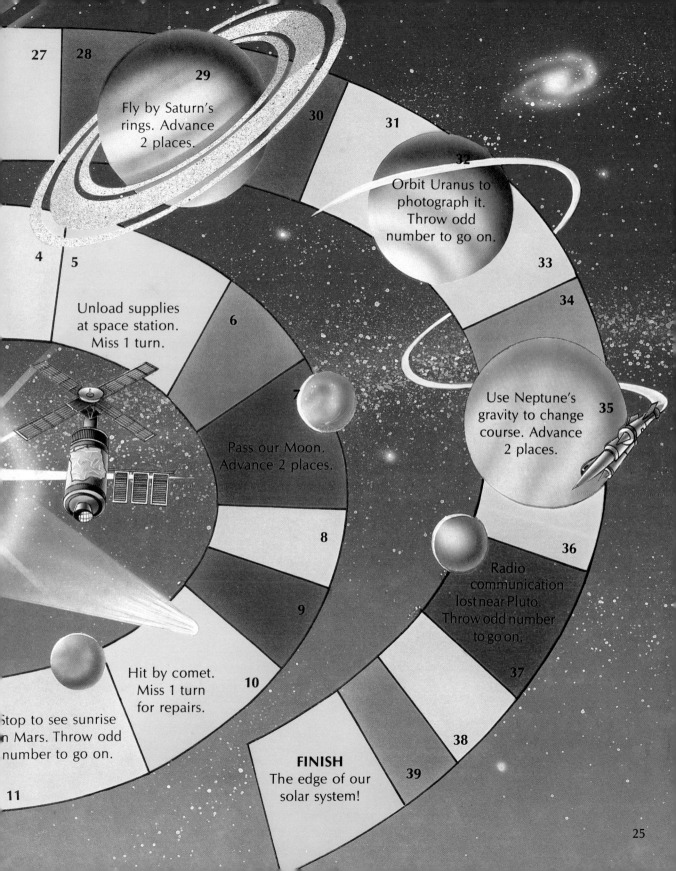

27

28

29
Fly by Saturn's rings. Advance 2 places.

30

31

32
Orbit Uranus to photograph it. Throw odd number to go on.

4

5
Unload supplies at space station. Miss 1 turn.

6

33

34

7

35
Use Neptune's gravity to change course. Advance 2 places.

Pass our Moon. Advance 2 places.

8

9

36
Radio communication lost near Pluto. Throw odd number to go on.

10
Hit by comet. Miss 1 turn for repairs.

37

Stop to see sunrise on Mars. Throw odd number to go on.

11

FINISH
The edge of our solar system!

39

38

Intergalactic Starship 1

Blip! Brrr! Hmmmmmm! An alien starship jumps into hyperspace and enters our solar system. Closer in, it is picked up by radar, and a message is pulsed out from Earth. This imaginary event is perfectly possible. Every star you see in the sky belongs to our galaxy. There are 100,000 million stars, and some may have planets on which life has evolved. What isn't possible is the starship's speed. Hyperspace, or travelling faster than light, is a science fiction speed.

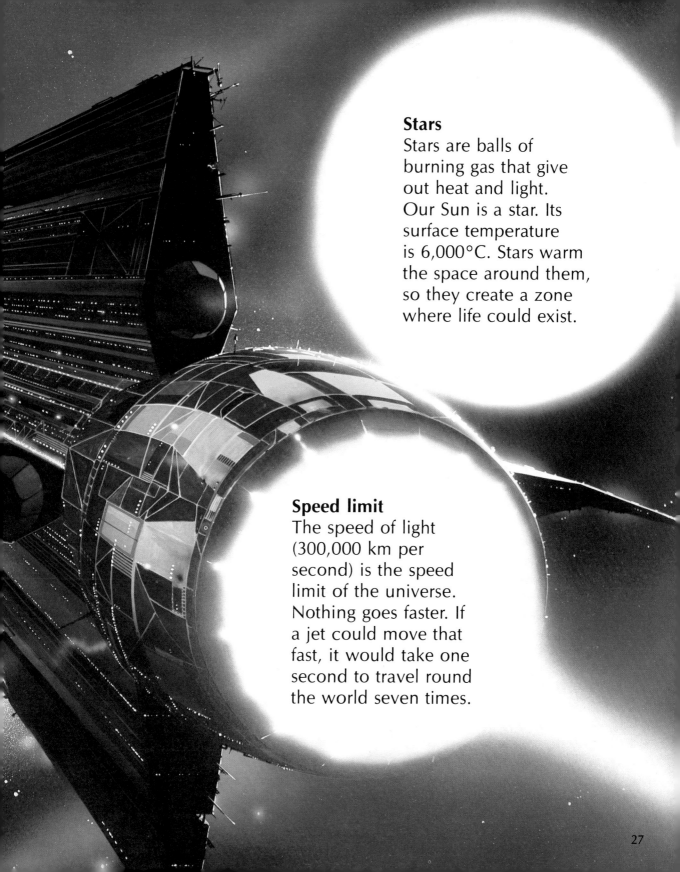

Stars

Stars are balls of burning gas that give out heat and light. Our Sun is a star. Its surface temperature is 6,000°C. Stars warm the space around them, so they create a zone where life could exist.

Speed limit

The speed of light (300,000 km per second) is the speed limit of the universe. Nothing goes faster. If a jet could move that fast, it would take one second to travel round the world seven times.

Message from Voyager 1

'X2TZ patrol ship to GOGOV space station control. We have intercepted an alien craft and are preparing to board. It's pretty battered, must have travelled from another galaxy.' An imaginary extra-terrestrial scout has found Voyager 1, a robot spacecraft launched from Earth in 1979. On Earth, the year is 41,999 AD.

The aliens are far less advanced than we are. Their robot craft is powered by nuclear generators! But they probably use solar power for travelling closer to their star.